Jobs People Do

Firefighters

by Mary Meinking

PEBBLE
a capstone imprint

Pebble Explore is published by Pebble, an imprint of Capstone.
1710 Roe Crest Drive
North Mankato, Minnesota 56003
www.capstonepub.com

Library of Congress Cataloging-in-Publication Data is available on the Library of Congress website.

ISBN: 978-1-9771-1375-7 (hardcover)
ISBN: 978-1-9771-1810-3 (paperback)
ISBN: 978-1-9771-1383-2 (eBook PDF)

Summary: Firefighters are the heroes we need in times of emergency. Learn all the roles firefighters have, the tools they use, and how people get this exciting, fast-paced job.

Editorial Credits
Gena Chester, editor; Kyle Grenz, designer; Jo Miller, media researcher; Kathy McColley, production specialist

Photo Credits
Alamy: AB Forces News Collection, 13; iStockphoto: Marco_Piunti, 4, Valerie Loiseleux, 25; Newscom: Ambient Images/Michael Routh, 23, Splash News/Ronald Asadorian, 28, ZUMA Press/Tass, 15; Shutterstock: Belish, 21, Dmitry Kalinovsky, 9, Drop of Light, 6, Eddie Gibson, 14, mat277, 17, Monkey Business Images, 19, Sergey Mironov, 1, sirtravelalot, Cover, Sundry Photography, 11, supergenijalac, 22, Ted Pendergast, 27, Toa55, 7

Printed in the United States 6589

Table of Contents

Words in **bold** are in the glossary.

What Is a Firefighter?

The scream of sirens fills the streets. Red lights flash. Cars pull over to the side of the road. Huge red fire trucks speed past. Firefighters have a job to do.

Being a firefighter is not easy. They are helpers. They are part of fire and **rescue** teams. They rescue people and pets. They put out fires in homes and in buildings. They risk their lives to save others.

Firefighters can be men or women. They care about others and want to help. They love their jobs.

Every day firefighters help people in danger. They must be brave. They rush into burning buildings. They work with police to save lives. They are some of the biggest heroes in the world.

What Firefighters Do

Firefighters work as a team. They fight fires. They use hoses with water or foam to put out fires.

Specially trained firefighters enter burning buildings. Their job is to find people. Others drive the fire trucks or use the hoses. The fire chief is the boss. He or she makes sure the job gets done and firefighters are safe.

Back at the station, firefighters get ready for the next **emergency**. They clean the trucks and set up gear. Firefighters are always ready to help.

Firefighters practice for all kinds of emergencies. They train to put out fires. They also train for medical emergencies. Training saves lives. Every second counts when people need help.

MARIN COUNTY FIRE DEPARTMENT
THROCKMORTON RIDGE
MARIN COUNTY FIRE DEPT.
1585

How Firefighters Help

The most important job for firefighters is to keep fires from starting. They teach fire safety. They tell us what to do if someone's clothing catches on fire. The person should stop, drop, and roll.

Fire Prevention Week is held every October. Firefighters teach people how to get away from a fire. They tell people to get low and go.

Firefighters help kids practice stop, drop, and roll.

Firefighters will help install **smoke alarms** in homes. They also tell people to check the alarms once a month. Every home needs smoke alarms.

Firefighters do more than put out fires. They help at car and plane crashes. Sometimes they help people after bad storms.

When firefighters get to an emergency, people may be scared or hurt. Firefighters are kind and try to help them.

Firefighters not only rescue people, they rescue animals. They help pets stuck in trees, on walls, or in water. Firefighters are good at fixing problems.

What Firefighters Use

Firefighters wear clothes that are **fire resistant** up to 1,200°F (648°C). Turnout pants help them dress quickly. Their pants are attached to boots. Firefighters jump into them. Then they pull on turnout jackets. Thick helmets and gloves protect them from heat and falling objects.

Firefighters carry air tanks. These are attached to full face masks. The masks help them breathe in smoky rooms.

helmet
mask
turnout coat
turnout pant
glove

Firefighters drive fire trucks. A pumper truck sucks water from a **hydrant** through a hose. The hose shoots water at a fire. Firefighters use axes or pry bars to break through doors or walls.

Thermal cameras are also needed. They show people's body heat. The cameras are used to find people in dark, smoky rooms.

ax
fire hose

Where Firefighters Work

Firefighters work all around the world. They work in small towns and big cities. They stop more than just house fires. They can work on ships and in airports. They work in mines and factories too. Firefighters are needed everywhere.

Some firefighters battle wildland fires. These firefighters sometimes work high aboveground. They use aircrafts to drop water or foam on huge fires.

A special airplane drops water on a fire

The best wildland firefighters are called hotshots. They fight fires that are hard to get to. They go whereever wildland fires start.

Smoke jumpers are firefighters who jump from airplanes. They land in the middle of a wildland fire. Their tools, food, and water are dropped to them. Smoke jumpers fight wildland fires in **remote** places.

How to Become a Firefighter

Do you want a job as a firefighter? People can go to college to study fire science. But college isn't always needed for this job. Most firefighters are **volunteers**. People who want to become firefighters must be at least 18 years old. They must pass written tests and health tests. **Professional** firefighters are paid to train and fight fires.

20
INSTRUCTOR

Only one out of every three firefighters is a professional. They went to a special school for firefighters. After finishing school, people work with **experienced** firefighters. They learn important skills on the job.

Firefighters are special people. They work in dangerous conditions. They often save lives.

DEPUTY
CHIEF
WALTHAM
WALTHAM
FIREFIGHTERS

Famous Firefighters

There are many famous firefighters. In 1736, Benjamin Franklin started Philadelphia's first fire company. Molly Williams became a firefighter in 1815. She was the first woman to do so. Steve Buscemi is a famous actor. He was also a firefighter. He helped New York City during the 9/11 attack.

Steve Buscemi

Fast Facts

- **What Firefighters Do:**
Firefighters put out fires.
They rescue people from danger.

- **Key Tools:**
fire hoses, axes, fire trucks, pry bars, thermal cameras, ladders

- **Key Clothing:**
fire helmet, turn-out pants and jacket, gloves, boots, mask, air tank

- **Education Needed:**
fire academy, medical training

- **Famous Firefighters:**
Benjamin Franklin, Molly Williams, Steve Buscemi

Glossary

emergency (i-MUHR-juhn-see)—a dangerous situation that must be dealt with quickly

experienced (ik-SPEER-ee-uhnsd)—skills or knowledge gained by actually doing something

fire resistant (FIRE ri-ZISS-tuhnt)—something that does not easily catch fire

hydrant (HYE-druhnt)—a large, upright pipe with a valve that draws water from a city's water supply

professional (pruh-FESH-uh-nuhl)—a person who makes money by doing something that others might do without pay

remote (ri-MOHT)—far away, by itself, or distant

rescue (RES-kyoo)—to save someone who is in danger

smoke alarm (SMOHK uh-LARM)—a machine that senses smoke and warns people by making a loud sound

thermal (THUR-muhl)—having to do with heat

volunteer (vol-uhn-TIHR)—to do something without pay

Read More

Charles, Audrey. *I Can Be a Firefighter*. New York: Gareth Stevens Publishing, 2017.

Liebman, Daniel. *I Want to Be a Firefighter*. Richmond Hill, Ontario, Canada: Firefly Books, 2018.

Staniford, Linda. *Firefighters to the Rescue Around the World*. Chicago: Heinemann Library, 2016.

Internet Sites

Firefighter Gear
https://www2.montgomerycountymd.gov/content/frs-safe/kids/firefighter_gear/firefighter_gear.pdf

Firefighting: Explain That Stuff
www.explainthatstuff.com/firefighting.html

NFPA Kids
www.sparky.org

Index